NABLE WORLD

ENERGY

Rob Bowden

KIDHAVEN
PRESS™

THOMSON

GALE

San Diego • Detroit • New York • San   aterville, Maine • London • Munich

For more information, contact
KidHaven Press
27500 Drake Rd.
Farmington Hills, MI 48331-3535
Or you can visit our Internet site at http://www.gale.com

Commissioning Editor: Victoria Brooker
Book Designer: Jane Hawkins

Book Editor: Victoria Brooker
Picture Research: Shelley Noronha, Glass Onion Pictures

Hodder Children's Books
A division of Hodder Headline Limited
338 Euston Road, London NW1 3BH

Cover: A wind-driven generator in Hawaii, USA, provides a sustainable source of energy.

Photo credits: Cover: Royalty Free/Corbis; title page Graham Kitching, Ecoscene; contents David Cumming/ Eye Ubiquitous; 4 Klaus Andrews/ Still Pictures; 5 Ecoscene; 6 Peter Hulme/ Ecoscene; 7 (top) Klaus Andrews/ Still Pictures, (bottom) Sally Morgan/ Ecoscene; 8 Brigitte Marcon-Bios/ Still Pictures; 9 Shehzad Noorani/ Still Pictures; 10 Mark Edwards/ Still Pictures; 11 Thomas Raupach/ Still Pictures; 12 Stephen Coyne/ Ecoscene; 13 Popperfoto; 14 Kevin Schafer/ Still Pictures; 15 (top) Sabine Vielmo/ Still Pictures; 15 (bottom) David Drain/ Still Pictures; 16 Carlos Guarita/ Still Pictures; 17 Hartmut Schwarzbach/ Still Pictures; 18 Toshiyuki Aizawa/ Popperfoto/Reuters; 19 Klaus Andrews/ Still Pictures; 20 Graham Kitching, Ecoscene; 21 Paul Gipe/ Still Pictures; 22 (top) Hartmut Schwarzbach/ Still Pictures; 22 (bottom) Mark Edwards/ Still Pictures; 23 Joerg Boethling/ Still Pictures; 24 Paul Glendell/Still Pictures; 25 Kevin Schafer/ Still Pictures; 26 Peter Hulme/ Ecoscene; 27 Martin Wright/ Still Pictures; 28 Mark Edwards/ Still Pictures; 29 (top) NASA/ Still Pictures; 29 (bottom) Mike Blake/ Popperfoto/Reuters; 30 Eriko Sugita/ Popperfoto/Reuters; 31 Kevin Nicol/ Eye Ubiquitous; 32 Ian Hodgson/ Popperfoto Reuters; 33 John Wilkinson/ Ecoscene; 34 Francois Gilson/ Still Pictures; 35 (top) Margaret McCarthy/ Still Pictures; 35 (bottom) Dylan Garcia/ Still Pictures; 36 Jorgen Schytte/ Still Pictures; 37 (top) Henning Christoph/DAS FOTOARCHIV; 37 (bottom) Joerg Boethling/ Still Pictures; 38 Martyn Chillmaid; 39 Alexandra Jones/ Ecoscene; 40 Rod Smith/ Ecoscene; 41 Dylan Garcia/ Still Pictures; 42 Ole Steen Hansen/ Hodder Wayland Picture Library; 43 Dorian Shaw/ Hodder Wayland Picture Library; 44 David Cumming/ Eye Ubiquitous; 45 (top) Julia Waterlow/ Eye Ubiquitous; 45 (bottom) Eriko Sugita/ Popperfoto/Reuters

LIBRARY OF CONGRESS CATALOGING-IN-PUBLICATION DATA

Bowden, Rob.
   Energy / by Rob Bowden
      p. cm.—(Sustainable world)
   Includes bibliographical references and index.
   ISBN 0-7377-1897-8 (lib. bdg. : alk. paper)
   1. Renewable energy sources. 2. Energy development I. Title. II. Sustainable
      World (Kidhaven Press)

   TJ808.B69 2004
   333.79'4—dc21

                                                                          2003052953

Printed in Hong Kong
10 9 8 7 6 5 4 3 2 1

Contents

Why sustainable energy?

ENERGY IS ESSENTIAL TO OUR SURVIVAL, but we are fortunate to live in a world surrounded by energy. We use energy every time we move; every time we breathe; even reading this book requires energy. We also use energy to power our vehicles and factories, and to heat and light our homes, schools, and offices. Our own energy comes from the foods we grow and eat, but other needs are met mainly by stores of energy, such as coal and oil, which were built up over millions of years. Today, we are using these stores very rapidly, much faster than they can be replaced or renewed. For this reason they are known as non-renewable energy sources.

Polluting coal-fired power stations provide energy that is far from sustainable and are harmful to both people and the environment.

ENERGY OVERLOAD

Non-renewable energy sources are becoming harder to find and more expensive to use. Of greater concern, however, is the fact that their use releases pollutants that cause great harm to the environment. Carbon dioxide (CO_2) is the most significant of these, as it is the main gas responsible for global warming. The level of CO_2 in our atmosphere is now 3 percent higher than it was before the industrial revolution of the nineteenth century and it is still rising. Most of the world's CO_2 currently comes from activities in more developed regions such as Europe and North America, but as the economies of less developed regions grow, their demand for energy is rising rapidly.

This wind farm on Lanzarote, one of the Canary Islands, provides a renewable source of energy.

If they rely on non-renewable energy sources to meet this demand then global warming is bound to speed up. The results could include climate change that ruins crops and causes regular flooding or drought, and sea level rises that will threaten cities such as Tokyo and London and whole countries including Bangladesh and several Pacific island states.

A SUSTAINABLE PATH

The world needs energy, and lots of it, but there are alternative sources of energy that do not pollute and will never run short. These sources include wind, water, and sunshine and they are all renewable energy sources. It is to these forms of energy that the world should increasingly turn if it is to follow a sustainable path, meeting the energy demands of its people today, but without harming the environment for future generations. One thing is for sure — most of today's energy supplies are far from sustainable!

DATABANK

In 1860 average world energy use was 0.9 MWh (megawatt hours).

By 1998 it had reached 19.1 MWh.

The energy problem

NEARLY ALL OF OUR ENERGY comes from a single source: the sun. Plants and algae convert the energy of the sun into a form of chemical energy through a process called photosynthesis. The energy is stored as carbohydrates or sugars and is the basic food that allows all life on earth to grow and function. This energy is passed on to other plants and animals (including humans) through the food chain.

A WORLD OF ENERGY

When plants or animals die they return energy to the earth as they decompose. Much of this energy is quickly taken up again by other plants in the form of nutrients, but over millions of years, enormous stores build up within the earth's surface. These stores are better known as fossil fuels and are found in the form of peat, coal, oil, or gas. Humans have developed methods of releasing and using the energy these contain for their own purposes. Today, fossil fuels are the main source of energy used to heat our homes, generate our electricity, fuel our cars, and power our industries.

This peat being harvested in Scotland contains the stored energy of dead plants and animals.

UPSETTING THE BALANCE

Human population has expanded from just 3 billion in 1960 to over 6.2 billion in 2002. At the same time the demand for energy has increased dramatically, more than doubling since 1950. Most of this demand has been met by the greater use of fossil fuels, but in doing so humans have now upset the balance of the earth's natural systems. As we burn fossil fuels to release their energy, CO_2 is released into the atmosphere in enormous quantities. Other gases, including nitrogen dioxide (NO_2) and sulphur dioxide (SO_2), are also released. As these mix with water vapor in the atmosphere they create an acid solution. This falls to earth as acid rain, killing trees and lakes and eroding the stonework of many buildings.

The effects of NO_2 and SO_2 are small, however, in comparison with the effect of CO_2 building up in our atmosphere. Scientists working in Antarctica have shown that current levels are higher than at any time in the last 420,000 years. They have also shown that by upsetting the balance of nature we are likely to experience climate changes as gases in the atmosphere form a blanket and trap heat close to the earth's surface.

The detail on this statue at Magdeburg Cathedral in Germany has been eroded by acid rain.

These icebergs were once part of a glacier in Torres del Paine National Park, Chile. Global warming is causing many glaciers to melt and break up.

LIFE ON A WARMER PLANET

The consequences of a warmer planet and climatic changes could be dramatic, and some scientists believe we are already experiencing these effects today. They may have a point: The 1990s was the warmest decade of the last millennium and 1998 was the warmest year since records began! One of the biggest effects of this warming is the melting of ice and glaciers in polar regions. Arctic ice sheets, covering the ocean, have reduced in thickness by 42 percent since the 1950s and some scientists believe that by 2050 they may melt completely during arctic summers. Mountain icecaps and glaciers are also melting as a result of temperature changes.

OPINION

Each one of us is responsible for global warming. Quite simply, it is the greatest environmental threat facing the world and it is happening at a rate faster than ever predicted.

The Ecologist "Go Mad: 365 ways to save the planet," 2001

Ice-melt is contributing to a rise in global sea levels by about one fourth of an inch per year. Sea levels in 2000 were already up to 7 inches higher than in 1900, but could increase by anything between 15–40 inches before 2100! A 3-foot rise in sea levels would flood almost a quarter of Bangladesh, affecting an estimated eighty million people. Sea levels are also rising as their waters warm, because warm water is less dense than cold water and so takes up greater space.

Ocean temperatures and currents influence wind directions and the amount of moisture in the atmosphere. In turn these affect our weather and so, as oceans warm, we are likely to experience abnormal weather patterns. In many parts of the world such abnormalities are already being reported such as torrential rains flooding normally dry regions, and long periods of drought striking areas that are usually well-watered. Extreme weather events such as cyclones are also on the increase and weather-related damages in the 1990s were five times higher than in the 1980s. Changes in the world's weather could have dramatic impacts on world food production, as crops are damaged or growing conditions change. Bangladesh alone stands to lose half of its rice land if sea levels rise as expected. Unfortunately the worst effects are likely to be felt in less developed countries, whose agricultural systems are already under pressure to feed their growing populations.

Factory workers commute to work during the 1998 floods in Bangladesh. Global warming and climate change could make such practices normal in the future.

DATABANK

The 23 warmest years, since records began in 1866, have all been since 1975.

Earth Policy Institute 2001

Women in Cameroon carry fuel wood back to their homes. Improving energy options for millions of people like them is a major challenge.

THERE IS AN ALTERNATIVE!

If we continue to depend on fossil fuels to meet our energy demands then there is every chance that the predictions of life on a warmer planet may come true. Indeed, some scientists believe that even if we were to stop using fossil fuels immediately, many of these effects would still be felt. But energy is essential for the running of our economies and daily lives, and in less developed areas of the world there are an estimated two billion people who do not yet have access to modern sources of energy like electricity. Providing energy to such areas is essential to improving their economies and helping to reduce poverty.

The main focus, then, is on how less-developed regions choose to develop their energy needs and what developed regions can do to reduce their reliance on fossil fuels. There are examples from across the world that show there are alternative ways to provide energy, without further harming the environment. These alternatives are known as sustainable energy or renewable sources and their use is growing rapidly as people become aware of the dangers of ignoring the energy problem.

WHAT IS SUSTAINABLE ENERGY?

Sustainable energy is energy that meets the needs of today's population without harming the environment or reducing the ability of future generations to meet their energy needs. To meet this goal sustainable energy is focused on renewable sources of energy such as the sun, wind, and water. This type of energy can be difficult to harness, however, and in some circumstances is not as reliable as traditional sources like oil or coal. Some solar power systems will only work efficiently in clear skies, and generating electricity from wind power requires steady winds of a specific speed. Harnessing sustainable energy sources is also still relatively expensive compared to using non-renewable sources. However, if the environmental costs of non-renewable energy are included, then sustainable sources are considerably cheaper by comparison as they have less impact on people and the environment. The challenge ahead is to make people more aware of their energy choices and encourage them to choose the sustainable option. This will demand a combination of technological and social changes over the coming years. This book will consider what lies ahead for energy in a more sustainable world.

As demand for sustainable energy increases, the cost of producing technology, such as these solar panels in Germany, will fall. Their low environmental impact means they are already cheaper than many other forms of energy.

weblinks

For more information on global warming go to www.epa.gov/globalwarming/kids

Developing sustainable energy

THERE IS, IN FACT, LITTLE NEW about harnessing sustainable forms of energy. Long before the use of coal or oil, people were making use of renewable and sustainable energy sources. The sun's energy, for example, has long been used as a means of drying food items in order to preserve them. Wind energy also has a long history and the first windmills date back to the seventh century in Persia (modern day Iran). Biomass fuels such as wood and dried vegetation date back thousands of years and are sustainable providing they re-grow faster than they are used.

In Germany, old and new windmills are both used to harness wind as an energy source.

LESSONS FROM THE PAST

In the past, energy use was limited by what was available in the surrounding environment and, together with much lower populations than today, this meant that energy use was largely sustainable. People understood the limits of their environment.

WARNING BELLS

Since coal was first used to power the industrial revolution, the global economy has depended on fossil fuels to meet most of its energy needs. By 1997 coal, oil, and gas accounted for 88 percent of the world's energy supplies, with most of this

In 1973, the OPEC oil crisis caused gasoline rationing and long lines at gas stations in the United Kingdom. This made people begin to question energy use and dependence on oil.

DATABANK

The first United Nations (UN) environment conference was held in Stockholm, Sweden, in 1972, and led to the founding of the UN Environment Program (UNEP), the global agency responsible for environmental monitoring.

(40 percent) coming from oil. In the 1970s, however, a number of incidents occurred that sounded warning bells about the continued dependence on fossil fuels. In 1973, a group of countries who were members of OPEC (Organization of Petroleum Exporting Countries) reduced their production of oil causing a sudden, fourfold increase in its price. This warned countries that they could not rely on a few suppliers of oil to meet their energy demands.

At around the same time, a number of reports were released looking at the impact of human actions on the environment. In particular, people started to criticize the use of fossil fuels as a source of energy because of the pollutants they released into the atmosphere. Others suggested that fossil fuels would run out before the end of the century, but we now know this not to be true. At the time, though, this combination of events led to something of an energy crisis and the search was on for alternative ways to meet the world's future energy needs.

This pipeline stretches almost 800 miles across the frozen wilderness of Alaska. It was built by the United States between 1975 and 1977 to transport oil from the remote Alaskan oil fields of the north.

RETHINKING ENERGY

The energy crisis caused industries, governments, and even individuals to rethink their energy needs. For some, the answer was conservation and a focus on energy efficiency. For example, aluminium production techniques in 2000 used about 25 percent less energy per ton than in 1980. Others focused on finding their own sources of energy. The United Kingdom developed its North Sea oil fields from 1973 onward. Countries without their own fossil fuels, or with only limited supplies, such as Kenya or France, turned to alternative energy sources such as nuclear and hydro electric power.

THE NUCLEAR OPTION

Nuclear power uses the energy released when atoms of uranium or plutonium are split to generate electricity. The first nuclear power station opened in 1954 in Russia, but nuclear power only became a major global energy contributor following the energy crisis. Between 1973 and 1990, energy production from nuclear power stations increased by 628 percent. The French government was amongst its most enthusiastic supporters and today nuclear power produces 75 percent of France's electricity.

This sign warns people about contaminated land. Land contaminated by nuclear waste can remain unsafe for hundreds, even thousands, of years.

Many welcomed nuclear power as the energy of the future; a cheap source of electricity with none of the atmospheric polluting emissions of fossil fuels. Then, in 1986, a nuclear power station at Chernobyl in Ukraine exploded spreading a radioactive cloud across much of Western Europe. This incident high-lighted the potential danger of nuclear power and many governments have since abandoned nuclear energy. Between 1990 and 2000 energy from nuclear power production grew by just 6 percent and only a few Asian countries have serious plans to continue developing nuclear power in the future.

Nuclear power creates few pollutants, but produces large quantities of hazardous radioactive waste.

DATABANK

Some of the waste products from the generation of electricity using nuclear power remain radioactive and dangerous for over 300,000 years.

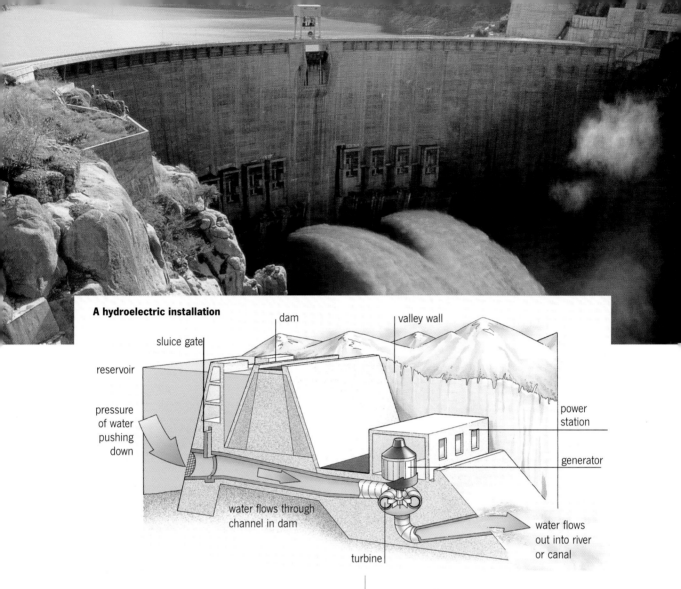

A hydroelectric installation

reservoir

sluice gate

dam

valley wall

pressure
of water
pushing
down

power
station

generator

water flows through
channel in dam

turbine

water flows
out into river
or canal

THE POWER OF WATER

Top: The Cabora Bassa Dam produces
renewable HEP for Mozambique.
Above: Diagram of HEP process.

As water falls it releases energy that can be captured and used to generate electricity. This is known as hydroelectric power (HEP) and is a clean and renewable form of energy. HEP normally involves the construction of large dams across river valleys and the creation of an artificial reservoir behind them. The water is then released through turbines that spin to generate electricity. Many countries have developed HEP to meet their energy needs. Kenya meets 67 percent of its electricity supplies in this way; in neighboring Uganda HEP accounts for 99 percent.

Although HEP provides a renewable form of energy it has many problems associated with it. Large areas of land are normally flooded by the reservoirs, often forcing millions of people to move as a result. The Three Gorges Dam, on the Yangtze River in China for example, will displace over a million people when it is finished.

Dams disrupt the natural flow of rivers and thus many rivers and reservoirs suffer from a build up of sediment behind dam

weblinks

For more information about hydroelectric power, go to wwwga.usgs.gov/edu/hyhowworks.html

walls. In the long term this may threaten the generation of electricity as turbines become blocked. Many experts now believe that large dams no longer provide a sustainable solution to energy needs, and for countries without suitable rivers they may not even be an option.

The Three Gorges Dam being built across the Yangtze River in China will not be finished until 2009. It will provide cleaner energy for China's ever growing energy demands.

A NEW DIRECTION

In the 1990s, growing concern about the environment, and particularly global warming, led to a renewed focus on energy issues. In 1992, world leaders met in Rio de Janeiro, Brazil, for the Earth Summit. Climate change was high on their agenda and 160 nations signed an agreement to tackle the issue. At a follow-up meeting in Kyoto, Japan, in 1997, targets were set to reduce emissions of greenhouse gases by 5.2 percent below their 1990 levels by 2012. Attention since Kyoto has focused on CO_2 emissions from the burning of fossil fuels and many countries are now following a new direction towards a future built on sustainable energy.

Delegates from around the world met in Kyoto, Japan, in 1997 to discuss reductions in the emission of gases that cause global warming. Energy production is the main source of such gases.

TECHNOLOGICAL IMPROVEMENTS

The incentive to reduce CO_2 emissions has led to a boom in sustainable energy technology. During the 1990s great improvements were made in the various technologies available and many countries now produce electricity at a similar cost to more traditional sources. For example, better knowledge about where to site wind turbines and increases in their size and efficiency has helped to reduce the cost of wind power by over

70 percent between 1980 and 2000. Continued technological improvements are expected to reduce the cost by a further 45 percent by 2015, making wind energy one of the cheapest energy sources available.

CHANGING ATTITUDES

Although technological improvements help to make sustainable energy more competitive, the changing attitudes of governments, industry, and individuals are even more important. As people begin to demand sustainable energy they encourage energy providers to switch toward sustainable solutions. In the late 1990s, several international oil companies realized that if they wanted to be the energy providers of the future they must begin investing in sustainable energy. Most now have a department specializing in renewable and sustainable energy solutions and some have become major forces in advancing sustainable energy. BP Solar, for example, is one of the world's largest producers of solar panels.

Modern wind turbines are more powerful and more efficient than earlier models. They are generally bigger, however, which means some people object to the sight of them.

---

weblinks

For more information about the Kyoto Protocol, go to
http://unfccc.iht/resource/convkp.html

Sustainable energy in practice

ALTHOUGH, GLOBALLY, SUSTAINABLE ENERGY accounts for just 2 percent of energy production, this figure is considerably higher in certain countries. These countries could be considered the pioneers of sustainable energy. Among the most impressive of the pioneers are Denmark, Germany, and Spain, all of which have experienced a rapid boom in sustainable energy (mainly wind power) in recent years.

This more unusual style of wind turbine is used to generate electricity in Alberta, Canada. Wind power has long been a pioneer of sustainable energy with many different turbine designs now in use.

THE PIONEERS

These pioneers are not alone, however. While other countries may be slower in developing sustainable energy, examples of such technology can now be found in virtually all corners of the world.

BLOWING IN THE WIND

The winds that circulate around the world are an inexhaustible supply of energy. Three states in the United States—North Dakota, Kansas, and Texas—receive enough wind to potentially provide electricity for the whole of the United States. Wind energy is captured using turbines that stand alone, or are arranged in clusters known as wind farms. The biggest modern turbines are around 260 feet tall and normally have two or three rotor blades that can be up to 115

Wind turbines loom over these houses in Germany. It is for this reason that many people object to having them located near to their homes. The development of off-shore wind turbines may provide a solution to this problem in the future.

feet long. The purpose of all turbines is to convert the motion of the rotor blades into electricity. This is done using a system of gears and a generator located at the top of the turbine tower inside a casing known as a nacelle. The electricity is then transferred in cables into the national grid or directly to where it is needed.

Wind energy is a clean and renewable form of energy and many countries are now actively placing wind at the center of their sustainable energy policies. Germany added almost as much wind power in 2000–01 as there was in the whole world in 1990. Other countries, including the United States, France, Argentina, China, and the United Kingdom, have recently announced ambitious plans to dramatically increase their share of energy from the wind. An area of particular growth has been that of offshore windfarms. Denmark, Germany, France, and the United Kingdom are among the European countries currently developing offshore windfarms. If these early projects prove successful then it is estimated that western Europe has enough offshore wind supply to power all of its electricity needs.

— **weblinks** —

For more information about wind power, go to http://telosnet.com/wind

Above: PV panels can be the most efficient form of providing electricity to remote locations such as this village in Sudan.

Left: PV cells can be used in many applications such as this solar-powered lawn mower in Sweden.

PV cells convert solar energy into electricity and are used in a wide variety of ways. The production of solar PV cells also increased over six-fold between 1990 and 2000.

SOLAR ENERGY

Solar energy is still a relatively expensive form of energy, but as solar technology improves and demand for sustainable energy increases it is becoming cheaper. Between 1980 and 2000 the price of photovoltaic (PV) cells fell by 83 percent.

Small PV cells are used to power wristwatches or calculators. Putting several larger cells together in a solar array can produce enough electricity for an average home. PV cells are a particularly useful form of sustainable energy as they can be installed wherever there is sunlight. This means they are often the cheapest form of electricity for areas that are not connected to an electricity grid. In South Africa, the government is aiming to install PV systems in 350,000 homes as part of its plan to bring electricity to poor rural areas. In neighboring Namibia PV cells are being used to provide energy for an innovative programme to connect remote schools to the internet.

PV cells also work in more cloudy countries — they do not need direct sunlight. In Norway, PV systems provide electricity for over fifty thousand rural homes and their use is growing annually. In 1998 Germany announced a plan to promote solar energy by installing PV systems on one hundred thousand rooftops. A similar program in Japan proved a great success following its introduction in 1997. By 2000, almost fifty-two thousand Japanese homes had had PV systems installed. In 1997, the United States started an ambitious plan to install a million PV rooftop systems by 2010, but in 2001 progress toward that target was slow.

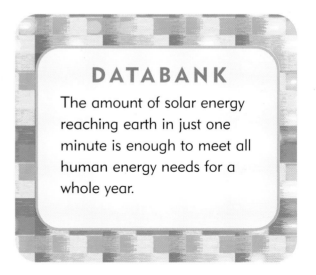

DATABANK

The amount of solar energy reaching earth in just one minute is enough to meet all human energy needs for a whole year.

The roofs of these homes in Hamburg, Germany, are covered with PV panels to provide their occupants with a sustainable and reliable source of energy.

Using solar energy to provide domestic hot water is now so easy that people can even install the systems themselves.

NON-PV SOLAR ENERGY

Converting solar energy into electricity using PV cells is the fastest growing use of solar energy, but there are other ways to use this plentiful energy source. In colder climates, one of the biggest uses of energy is the heating of buildings. By changing the design of buildings it is possible to use the heat of the sun, even during winter. Large south-facing windows allow solar energy to enter the building, and special building materials absorb the heat, releasing it slowly throughout the day. This use of passive solar energy can reduce heating bills by around 50 percent. Such buildings also reduce the need for artificial light, further saving energy.

Solar energy can also be used to heat water. Panels containing a liquid (normally placed on a rooftop) heat up in the sun. The liquid circulates around the building in pipes providing heat through radiators, or warming water in a storage tank for later use. In cooler and cloudier climates such systems can only heat water, but they can reduce the need for energy from other sources by about two-thirds.

On a much larger scale the sun's energy can be concentrated using hundreds of giant mirrors and reflected on to a tank or pipes containing a liquid — often oil. The superheated liquid (temperatures reach almost 752°F) is used to produce steam to power electrical generators. In the United States, nine such solar-thermal power stations in the Mojave Desert have been active since 1985.

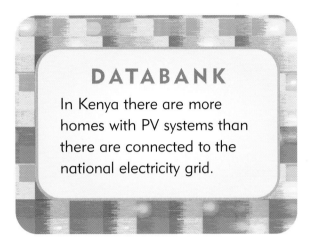

DATABANK

In Kenya there are more homes with PV systems than there are connected to the national electricity grid.

weblinks

For more information on solar energy go to www.solarenergy.org/kids.html

These fields of mirrors are part of the world's largest solar-thermal power station in the Mojave Desert, California.

GEOTHERMAL ENERGY

In some parts of the world, the intense heat at the center of the earth (reaching temperatures of over 8852°F) is found much closer to the surface as it rises through faults (cracks) or vents in the earth. The regions where this happens are normally associated with volcanic activity.

DATABANK

The U.S. Department of Energy estimates that potential geothermal energy is fifty thousand times greater than all of the world's oil and gas resources.

Water trapped within the rocks heats to extremely high temperatures and can be extracted to provide a potential form of energy called geothermal energy. Geothermal energy can be used to provide direct heat, by literally piping the heated water around homes, offices, and factories. In Iceland, around 85 percent of the population heat their homes in this way. The capital, Reykjavik, has one giant district heating system that meets the heating needs of almost 150,000 people. Similar systems exist in the United States and New Zealand among other countries.

This superheated steam is rising from the ground in Iceland and is captured to generate geothermal power — a renewable and sustainable form of energy.

weblinks

For more information about geothermal energy, go to http://geothermal.marin.org

Alternatively the superheated water or steam can be used to drive generators and produce electricity. Some twenty countries currently produce electricity in this way including New Zealand, the United States, Mexico, Japan, Australia, Kenya, and Iceland. Many others are now exploring the potential of geothermal energy as a clean and sustainable source for future energy needs.

BIOMASS ENERGY

Biomass, for example, wood and agricultural waste, is still the main source of energy for around a third of the world's population who burn it directly to obtain heat and some light. Although this is sustainable (providing trees and plants are replaced at the same rate as they are used) it produces limited amounts of energy and the collection of biomass can be very time consuming. In China, women can spend an average of three hours per day collecting biomass for energy.

Another use of biomass is to generate electricity. Biomass furnaces heat water and produce steam for turning electricity generators. Excess heat can also be used to provide heating for individual buildings or whole districts. In the United States the paper and pulp industry is a major user of this combined heat and power (CHP) technology, using the waste from paper production as the biomass fuel to generate electricity.

Wood is still the main source of energy for millions of poorer countries. This poster in Burkina Faso promotes the use of more efficient wood stoves to reduce deforestation rates.

Manure is added to a biogas plant in Andhra Pradesh, India. As it decomposes it will release methane gas that can be used as a source of energy.

BIOGAS

Biomass can also be turned into biogas by burning it in a chamber starved of oxygen. As biogas it can be used like natural gas for heating or generating electricity. China built over seventy such biogas plants during the 1990s, each serving up to sixteen hundred families with gas for cooking. In India, biogas is used for generating electricity and there are plans to build biogas plants that will use waste (bagasse) from India's sugar mills.

Methane is a naturally occurring biogas generated as biomass decomposes. In several countries methane is captured from landfill sites as biological waste decomposes and is used as biogas for generating electricity and/or heating. The United States alone has over two hundred such biogas systems. Using methane instead of allowing it to enter the atmosphere has an added benefit because it is one of the greenhouse gases responsible for global warming. In fact methane is about twenty-two times more damaging than CO_2 in terms of climate change.

THE HYDROGEN REVOLUTION

The most exciting energy development in recent years has been the potential of energy from hydrogen — the most abundant element in the universe. Hydrogen has long been used in certain industries and is one of the main fuels that powers rockets and space shuttles into orbit. The idea of a hydrogen revolution, however, follows the recent development of hydrogen fuel cells. These cells combine hydrogen with oxygen from the environment to create a chemical reaction that generates electricity, water and some heat. The electricity and heat can be used while the water is a harmless by-product.

Individual fuel cells can be combined into sets to produce vast amounts of electricity. At present, most interest in fuel cells lies in their potential as a replacement for gasoline or diesel engines as the means of powering vehicles. Several manufacturers have begun developing their own hydrogen-powered vehicles. BMW released a prototype in 2001 capable of 140 miles per hour, and aim to sell one thousand such vehicles by 2010. Daimler-Chrysler hope to release their hydrogen-powered Necar 4 by 2004 and other automotive manufacturers including Ford, General Motors, Honda, and Nissan are also developing hydrogen-powered vehicles.

The shuttle blasts off into space thanks to the power of hydrogen.

This bus in Burnaby, Canada, is one of the first to be powered by a hydrogen fuel-cell. With zero emissions, such technology could provide the sustainable energy of tomorrow.

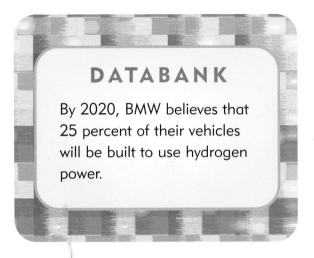

DATABANK

By 2020, BMW believes that 25 percent of their vehicles will be built to use hydrogen power.

THE HYDROGEN PROBLEM

The problem with hydrogen is that it is not a naturally occurring energy source like gas or wind, it has to be created using another form of energy. Creating hydrogen uses enormous quantities of energy, most of which comes from unsustainable fossil fuels. In addition most of the hydrogen is currently produced by extracting it from fossil fuels which are made of hydrogen and carbon. Made in this way, hydrogen is not a sustainable form of energy. In the future, however, hydrogen could be produced by splitting water into hydrogen and oxygen using a process called electrolysis. Although this uses large quantities of electricity, if it was done using sources such as solar or geothermal energy then hydrogen would be a truly sustainable energy.

Passersby inspect a new Honda car powered by a hydrogen fuel cell.

Countries such as the United Arab Emirates could soon be producing hydrogen using their plentiful supplies of sea water and solar energy.

Iceland has been quick to recognize the potential of hydrogen energy. It has announced plans to become the first hydrogen economy, using its vast supplies of geothermal energy to produce hydrogen from water. The Middle East is also poised to take advantage of the hydrogen revolution. With plentiful supplies of sunlight, the region is ideally suited to use solar energy for producing hydrogen from water. In fact, so good is the potential that in 2001 Saudi Arabia's energy minister (one of the founders of OPEC) declared that he could foresee the day when water would be more valuable than oil in the Middle East.

In the long term, hydrogen fuel cells could be used to generate clean energy, not just for vehicles, but for offices, schools, laptop computers, and even mobile phones. In fact, as the technology improves, we could soon have fuel cells in our homes with hydrogen piped to each house in much the same way as natural gas is in many regions today. One U.S. organization has predicted that the number of stationary hydrogen fuel cells for generating electricity will grow by almost two hundred times between 2001–2010. It must be remembered, though, that hydrogen energy is not truly sustainable unless it is produced by electrolyzing water using other sustainable energy sources.

Making sustainable energy work

A FUTURE IN WHICH SUSTAINABLE ENERGY plays a key role is about much more than just technology. In many ways, technology is the easy part of the energy problem. The greatest challenge is to change the way people think about and use energy in their lives. This is not an easy task. The long dependence on cheap and readily available fossil fuels means that many people take their energy needs for granted and may be resistant to changes, especially if it will cost them more. For example, in the United Kingdom a rise in gas and diesel prices led to country-wide protests in September 2000 that brought roads to a standstill and caused major fuel shortages. In this case the price rise was not due to sustainable energy policies, but it provided an indication of how people may react to such price changes in the future.

Fuel-cost protests in the United Kingdom in September 2000 led to panic buying as people feared supplies would run short and prices increase.

Visitors to the Centre for Alternative Technology, in Wales, can learn all about the benefits of using sustainable energy and see it in action.

MORE THAN TECHNOLOGY

Education is of great importance in encouraging the switch toward sustainable energy. Making people aware of the consequences of continued dependence on fossil fuels and of the many benefits of sustainable energy is vital to its success. People also need to be informed about the sustainable energy options available to them. While sustainable energy remains more expensive than traditional energy sources, many believe education is not enough. They argue that governments should intervene to make sustainable energy cheaper or alternatively make non-sustainable energy more expensive. Such interventions are known as price incentives and have been highly effective where they have been introduced, as we shall learn later. In reality, education and intervention are often most effective when they are used together.

weblinks

Find out how you can make your house more energy efficient. Go to www.eere.energy.gov/consumerinfo/energy-savers

THE REAL PRICE OF ENERGY

In debates about sustainable energy, one of the most frequent arguments that you will hear is that it is too expensive. If you look purely at the price paid for energy then you can see where this argument comes from. In Europe, electricity produced from coal costs less than half what electricity from wind energy costs. Electricity from solar energy is even more expensive, almost 20 times the cost of producing energy from coal.

Supporters of sustainable energy point out that many prices are unrealistic and do not represent the real price of energy. They argue that in calculating the real price of energy we should include the costs that using an energy source has on the environment, human health and the economy. These costs are known as external costs, or sometimes as social costs because they are costs that fall on the whole of society, whether or not they are users of that energy. If external costs are included then the price of energy can look very different, but such costs are very

Many people ignore the costs to the environment and human health when they think about the price they pay for different energy sources.

Above: Smog caused by vehicle emissions hangs over Los Angeles, California, like a choking blanket.

Right: This man uses an inhaler to treat his asthma. The number of people with asthma is thought to be increasing due to air pollution.

difficult to calculate. The European Union has attempted to do this in a study that lasted over 10 years (1991–2001). Their research suggests that the real cost of producing electricity from coal is up to over twice the price actually charged to customers. Wind energy on the other hand was shown to have very few external costs and would barely cost more than its actual cost today, and would be cheaper than coal.

Another factor to consider in calculating the real price of energy is the energy used in constructing the different options

available. In the United Kingdom, for example, an average wind farm will repay the energy used in its construction within five months. Over its lifetime, each wind turbine will produce around thirty times more energy than was used to make it. By contrast, coal and nuclear power stations produce only a third of the energy used to build and run them.

POLLUTER PAYS PRINCIPLE

Government support for technology such as this solar water heater installed in Bangalore, India, can do much to encourage the growth of sustainable energy use.

Some countries have introduced energy policies that charge more for use of polluting energy than for use of clean energy. This is known as the polluter pays principle. In Denmark, Norway and Sweden, the governments have introduced taxes on energy produced using fossil fuels. The Danish plan means that from 2003, 20 percent of electricity consumed must be produced from renewable sources. Electricity users will be given green certificates when they buy from renewable sources to prove to the government that they are meeting this target. In 2001, the United Kingdom also introduced a form of energy tax called the climate change levy. Businesses and industries are charged a tax

(levy) on their consumption of electricity, to take account of the climate change caused by using that energy. As part of its policy to reduce emissions of CO_2, the United Kingdom government allowed energy from renewable sources to be exempt from the climate change levy. This acts as an incentive for businesses to look for energy from more sustainable sources.

SUBSIDIES FOR SUSTAINABILITY

Taxes act as a form of fine for the continued use of non-sustainable energy. The higher cost (because of the tax) is intended to persuade people to switch over to more sustainable sources. However, in many countries, sustainable sources are not yet developed enough and so people simply end up paying the tax. Experts believe that to encourage the faster development of sustainable energy, governments must offer some form of financial assistance. When financial help is given in this way it is known as a subsidy.

In Japan, the government has been subsidizing the installation of PV systems since 1994. The government pays around a third of the cost of installation and in return the user provides regular reports on their use of the system, including any problems they have had. This information helps the government to further improve the program. In Germany, the government provides a ten-year interest-free loan to those who install solar systems. In the United States, a reduction in the amount of tax paid (a tax credit) on the production of wind energy was responsible for the rapid growth of U.S. wind power in the late 1990s. The tax credit plan ended in December 2001, but there are hopes it will soon be extended. In India, various subsidies have helped its sustainable energy program become one of the biggest in the world. Many point to India's success as a model for meeting the future energy needs of developing countries.

The roof of this house in Osaka, Japan, is covered in PV panels. The Japanese government supports the use of PV roofs by providing grants and loans.

Solar lamps such as these being mended in Rajasthan, India, can greatly improve the lives of rural villagers without harming the environment.

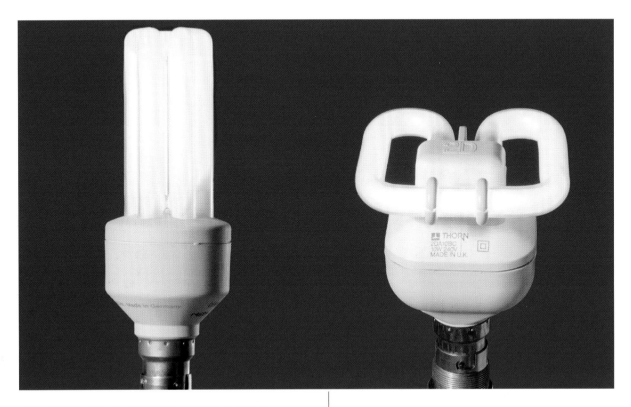

ENERGY EFFICIENCY

Compact Fluorescent Lamps (CFLs) can greatly reduce energy use, without reducing the amount of light produced. They can also save you money!

Taxes such as the United Kingdom's climate change levy, which charge according to the amount of energy used, promote greater efficiency in energy use. Energy efficiency is an important part of promoting sustainable energy because it can be achieved immediately, and often at very little, or indeed no cost. In fact in the long term, most forms of energy efficiency will save money through lower energy bills, even if their initial cost may seem expensive. The best example of this is the compact fluorescent lamp (CFL) — a simple energy efficient light bulb. Although CFLs cost several times more than normal light bulbs, they last around ten times longer and use 75 percent less energy to produce the same light. In Denmark a CFL used for four hours a day will save enough energy to pay back its purchase cost in just six months. Over its ten thousand–hour lifetime it will save energy costs equivalent to twelve times its purchase price. In 1999 the estimated 1.3 billion CFLs in use worldwide, saved the energy (and emissions) equivalent of twenty-eight coal-fired power stations.

Energy efficiency is possible in most situations. Danish scientists for example have adapted Coca-Cola bottle coolers (refrigerators like those you find in convenience stores) by installing more efficient insulating glass and new low energy cooling mechanisms. This has reduced their energy consumption by 40 percent. Forty of the new bottle coolers were successfully introduced into stores in the Copenhagen and Aarhus areas during 1999. With an estimated seventy thousand bottle coolers in Denmark alone, and millions more worldwide, the potential energy savings are enormous.

In the United States an Energy Star labeling system was introduced in 1992 to inform consumers about energy efficient goods, even homes. Those that meet

---
weblinks

For more information on the Energy Star program, go to www.energystar.com

---

energy saving targets are awarded the star. As consumers begin to purchase such products they will send signals to other manufacturers to improve the efficiency of their goods too.

A labeling program in Australia helps consumers choose energy efficient appliances that are better for the environment.

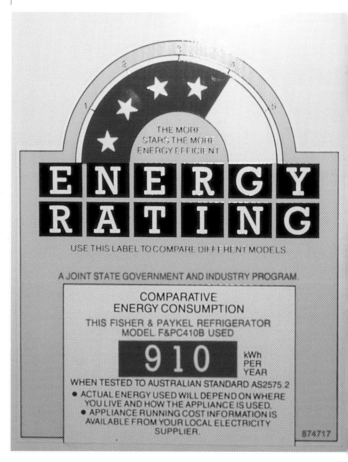

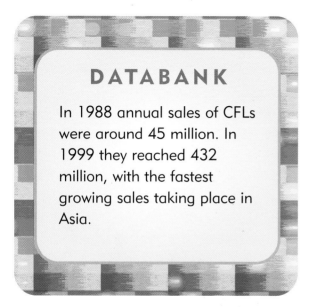

DATABANK

In 1988 annual sales of CFLs were around 45 million. In 1999 they reached 432 million, with the fastest growing sales taking place in Asia.

Sustainable energy and you

∙∙∙

A KEY TARGET FOR THOSE CONCERNED with sustainable energy is encouraging greater awareness of the issues involved. Education, at all ages, plays a lead role in this, but making young people aware is especially beneficial. This is because you will then be able to make informed choices throughout your lifetime. Governments and campaigners in most countries are now producing information about sustainable energy for young people. The Internet in particular has lots of useful websites where you can learn more.

EDUCATION AND AWARENESS

∙∙∙∙∙∙∙∙∙∙∙∙∙∙∙∙∙∙∙∙∙∙∙∙∙∙∙∙∙∙∙∙∙∙

In some countries special centers have been set up to promote and teach about sustainable energy. One of the oldest of these, the Centre for Alternative Technology (CAT) is located in Wales in the United Kingdom. Founded in 1973, CAT is a living and working community powered by various forms of renewable energy sources. Visitors to CAT can learn about how the different technologies work, with many hands-on practical

An environment shop in Hull, in the United Kingdom, provides people with information about what they can do to make their lives more sustainable.

The Centre for Alternative Technology in Wales provides courses and information on how to adopt a more sustainable approach to living.

examples to try. There are similar education centers opening throughout the world. You could learn a great deal by finding out if there is one near to you. Visit with your family, or suggest a school trip for your classmates.

DIRECT ACTION

There is much you can do right now to change your energy consumption and contribute toward a sustainable future. Find out, for example, whether your home or school is using CFLs and if they are not then explain why they should. Even if you already use CFLs, remembering to turn them, and other electrical appliances, off when not in use can save a great deal of energy. For example, a video recorder left in standby mode uses almost as much electricity as when it is playing! It has been estimated that in the United Kingdom, television sets and videos left in standby mode use $250 million worth of electricity over a year. There are, in fact, hundreds of simple behavioral changes that we could all make in our daily lives that would save energy.

— weblinks —

For more information on the Centre for Alternative Technology, go to www.cat.org.uk

The most efficient way to use a dishwasher is when it is completely full.

LOCAL ACTION
Save fuel, save resources

- Put on an extra layer of clothing instead of turning on the heat.

- Walk or cycle for short journeys instead of using a car.

- Make the room with most natural light your family's main living area.

- Keep energy trapped by closing doors and windows and blocking drafts.

- Only use a dishwasher or washing machine when it is full.

- Recycle goods and possessions to save the energy needed to make new ones.

If we all began to make simple changes such as these then we would not only reduce energy consumption and emissions, we would also save money! You can add to this list yourself and trade ideas with others.

PURCHASING POWER!

By thinking carefully about what we buy, consumers like you and me can put great pressure on businesses to switch toward sustainable energy. We have already learned about the Energy Star labeling system in the United States. Similar programs now exist in many countries, so next time you purchase an energy consuming product ask about its energy efficiency.

In most countries you can now also buy your electricity from suppliers who actively support the use of sustainable energy. Some companies offer what is called an eco-fund tariff (tax). This is where you pay a little extra for your electricity and the company use this money to invest in developing new sources of sustainable energy. A green tariff or renewable tariff is also offered by some companies and assures that all of your electricity is generated from sustainable sources.

OPINION

Every time we use electricity to switch on a light, or watch television, C02 emissions are released into the atmosphere. Purchasing Green electricity would avoid this.

The National Energy Foundation, UK

In the United States, an eco-fund tax in Los Angeles charges customers around 6 percent more for their electricity. In return the local provider (Los Angeles Department of Water and Power (LADWP) supplies 20 percent of their electricity from sustainable sources and invests in creating further such sources for the future. Customers also receive assistance with energy saving advice for their home. In 2000, LADWP had signed up twenty thousand customers for the program, around 1.5 percent of its total customers. They plan to increase this to two hundred thousand customers before 2005. The experience of LADWP shows that people are ready to make choices and switch to a future powered by sustainable energy.

Getting out and about on bicycles not only reduces energy use, but is good for your health and is enjoyable, too.

The future of sustainable energy

WE HAVE SEEN IN THIS BOOK THAT although the world is faced with serious energy challenges, there are many instances where a positive start has been made to meet them. Countries such as Denmark, Germany, and India are leading the way toward a future built around sustainable energy, and others are catching up rapidly. Even so, there is still a great deal to do.

This bus has been adapted to educate people about the benefits of using solar energy. It travels around India and demonstrates the different technologies available to people. Promoting sustainable energy in this way is an important task.

THE CHALLENGE AHEAD

Global energy demand is expected to grow by 57 percent over the period 1997–2020 and energy forecasts suggest that fossil fuels will still account for 90 percent of all energy in 2020. Oil will remain the dominant energy source accounting for about 40 percent of use, most of it for transportation. Natural gas (26 percent) will replace coal (24 percent) in second place and nuclear (5 percent) and HEP (3 percent) will follow. Sustainable energy sources are expected to grow faster than any other form of energy, but will still make up the smallest proportion of world energy supply, providing just 3 percent by 2020.

As the economies of less developed countries grow they will demand ever-greater supplies of energy. This new Pudong business and industrial district is on the edge of Shanghai in China.

Less developed countries will account for over two-thirds of the increase in energy demand between 1997–2020. This is partly due to continued population growth in these regions, but also because their economies are still industrializing. In more developed regions industrialization required large amounts of energy and led to many of today's environmental concerns. The global challenge then is how less developed regions can industrialize and develop their economies without producing more of the problems that have accompanied industrialization in the past. Central to this challenge will be the greater use of sustainable energy sources.

PERSONAL CHOICE

Governments and global organizations can not be relied on to make the switch to sustainable energy on our behalf. We must make choices, too. There are already sustainable energy options available and they are growing daily. If we choose to ignore them and the energy problem then we must be prepared to live with the potential consequences. If we choose to take action however, we will have begun to make a difference; a difference that will help build a more sustainable world.

Children from Kyoto, Japan, hold a giant silk Earth to celebrate the launch of the Kyoto conference to tackle global warming. Decisions made today will affect their future and the future of their children.

Glossary

Acid rain Produced when pollutants, such as sulphur dioxide and nitrogen oxides (emitted when fossil fuels are burned), mix with water vapor in the air

Alternative energy A term often used in place of renewable energy or sustainable energy to describe alternative sources of energy to the fossil fuels that dominate current world energy use.

Biomass fuels Consisting of biological matter (wood, crop stalks, animal dung, and collected leaves), or methane gas made from decomposing biological matter.

CFL (Compact Fluorescent Lamp) Energy-efficient light bulb that lasts around ten times longer and uses 75 percent less energy than a standard light bulb.

Climate change levy Energy tax introduced by the U.K. government in 2001. All businesses/industries must pay a tax on the amount of electricity they consume to compensate for the climate change caused by using that energy.

Developed countries The wealthier countries of the world including Europe, North America, Japan, Australia, and New Zealand.

Eco-fund tariffs Consumers pay an extra tax for their energy to support investment into sustainable energy sources.

Fossil fuels Fuels from the fossilized remains of plants and animals formed over millions of years. They include coal, oil, and natural gas.

Geothermal power The use of superheated steam from deep underground to drive turbines for generating electricity.

Global warming The gradual warming of the earth's atmosphere as a result of greenhouse gases, such as carbon dioxide and methane, trapping heat.

Greenhouse gases Atmospheric gases, such as carbon dioxide and methane, that trap some of the heat radiating from the earth's surface. Greenhouse gases contribute to global warming and climate change.

Hydroelectric power (HEP) Electricity generated by water as it passes through turbines. HEP involves damming river valleys and forming artificial lakes.

Hydrogen energy Hydrogen and oxygen combine to cause a chemical reaction in which electricity, heat and water are created.

Industrial Revolution The period in the late eighteenth and early nineteenth century when new machinery, and the use of fossil fuels to generate energy, led to the start of modern industry.

Megawatt (MW) A measure of electrical power most often used to describe the power output of electricity sources such as power stations or wind farms. One megawatt is equal to one million watts.

National grid Network of power lines that transmits electricity from power stations and distribution centers to homes and businesses.

Non-renewable energy sources Energy sources, such as coal, oil, and natural gas, that once used cannot be replaced.

Nuclear power Electricity produced when atoms of uranium or plutonium are split.

For further exploration

Photosynthesis Process whereby plants, using the sun's energy, convert carbon dioxide and water into carbohydrates (used for plant growth) and oxygen.

Pollutants Any substance that pollutes another. For example, chemicals and waste products that contaminate the air, soil, and water.

PV cells Photovoltaic cells that convert the sun's energy into an electrical current.

Radioactive Substances, such as uranium or plutonium, that emit energy in the form of streams of radioactive particles. These particles are extremely harmful to humans and animals if they come into direct contact with them.

Renewable energy Sources of energy that are continually renewed and will never run out, such as the sun, the wind, and waves.

Solar power Electricity generated by converting energy from the sun, normally using solar panels of PV cells.

Subsidy A payment, normally made by governments, to encourage certain practices.

Sustainable energy Energy that meets the needs of the present generation without harming the environment, or the ability to generate energy, for future generations.

Turbines Machines consisting of rotor blades which turn under a force (i.e. water, wind) to generate electricity.

Wind power Electricity generated from the wind. Turbines capture wind energy with moving rotor blades and convert it to electricity.

BOOKS

Ian Graham, *Energy Forever?: Geothermal and Bio-energy*. London: Hodder Wayland, 2001.

Ewan McLeish, *21st Century Debates: Energy* by London: Hodder Wayland, 2001.

Robert Snedden, *Essential Energy: Energy Alternatives*. Barrington, IL: Heinemann Library, 2001.

Index